Caleb Harlan

Elflora of the Susquehanna

Caleb Harlan

Elflora of the Susquehanna

ISBN/EAN: 9783337270247

Printed in Europe, USA, Canada, Australia, Japan

Cover: Foto ©berggeist007 / pixelio.de

More available books at **www.hansebooks.com**

OF THE

SUSQUEHANNA.

A POEM.

BY
C. HARLAN, M.D.

PHILADELPHIA:
PRINTED FOR THE AUTHOR,
BY J. B. LIPPINCOTT & CO.
1879.

PREFACE.

The following Poem, excepting a few verses, was written more than forty years ago.

Why was it not then published? The reason was, the author would not spare the means, at that early day, to bring out the work in a handsome and attractive form. Hence it was laid away and the time of its appearance indefinitely postponed. But now my gray hairs remind me that ere long I shall be called upon to experience the grandest event in human life—the departure of my soul to a more glorious home—and that it is time to put my affairs in order for the change. Large piles of letters and papers must be burned, or disposed of in some other way. And at last I must decide what I will do with the manuscript of ELFLORA OF THE SUSQUEHANNA.

It must be destroyed or published. I cannot bear to leave it to the tender mercy of some disinterested person. It might fall into the hands of a literary pirate, or it might be brought out in a garbled edition—a disgrace to the author and an insult to the heroine. Then burn the poem! No, I will not. It shall be published, out of regard to that beautiful and pure being whose history I have attempted to portray.

The scenes, the characters, interested my boyhood. The events were engraven upon my heart. I loved them as we cherish the idols of other days. Then why not preserve them in poetry? Yes, ELFLORA! for thy sake alone I endeavored well to do it. Thou wert ever fond of the voice of the Muse, and, though the world may neglect thee, in the homes and the hearts of a few, if possible, thou shalt be enshrined for ever.

Elflora of the Susquehanna.

CANTO THE FIRST.

I.

Though cool the morns, though fresh the winds at play,
Though mild the warmth that blooms the blush of May,
Throughout all Nature hath a change begun,
The forest shades proclaim the Spring hath come;
The watchful birds December winds had driven
To seek in torrid lands a snowless heaven,
With cheerful song now hail the halcyon hours,
A home shall greet them wreathed in fruits and flowers.

II.

By Susquehanna's waves, that brightly sweep
Her hundred hills enlightened freemen keep,
Secluded stands a mansion, mouldering on
In shade and moss through shining seasons gone.
The verdant lawns, expanding far and wide,
The groves, the vales, that fringe the crystal tide,
The distant forests, piled along the sky,
Majestic oaks, in opening vistas nigh,
The wooded highlands, capt with granite towers,
The gushing fountains, wildering greenwood bowers
Around the dwelling,—ever please the eye
With rural Nature's rich variety.
And in that mansion—in the library hall,
Where old engravings crown the faded wall,
Where books and paintings, pebbles rare, and flowers,
Betray the ramblings far, the well-spent hours—

CANTO THE FIRST.

ELFLORA sits alone. This spring-like day
She seems amused with some old English lay;
But now, attracted to the window nigh,
She looks intently on the clearing sky—
On white-blue mist just mantling o'er the vale,
On dazzling dewdrops loosed by breeze nor gale,
On river sparkling, as it passed in view,
With stars the sunlight o'er its surface threw,
And, roused and listening, every feature flush'd,
Her joyous thoughts in fervent language gush'd:
"O brilliant hour! enrapturing morn!
Upon thy fragrant airs are sweetly borne
Birds' warbling melodies; and, soft and clear,
The purling of the fountains greets my ear.
Ye little fragments of a broken storm
That pass along the sky, and, changing, form
A thousand shapes that catch the burning ray,
And shining, silver-fringed, float far away!

Oh, see ye not yourselves in yonder stream
Now mirrored bright in morning's hallowed beam?
Those placid waters image back the sky,
The beetling crags, the deep-green laurels nigh;
And this dark forest, now in sombre gloom,
Which is to me more dear than gay saloon,
Unrivalled beauties paint its vernal shades,
Wave o'er the cliffs, and deck its sunless glades,
Invite the weary to its peaceful groves,
Instruct the curious and the sad compose.
Though ever lovely, I rejoice to see
Its wildwood blossoms bloom once more for me."

III.

'Twas thus ELFLORA spoke; then from the door
She lightly stepped, array'd to seek the shore.
The lawn she cross'd, the craggy steep descends,
Through laurel winds, and treads the deepening glens

CANTO THE FIRST.

Where towering oak and lofty ash unite
Their clouding limbs that half exclude the light—
Where wary huntsmen find their noblest game—
Where prowling wolf and wild deer oft are slain,
She wanders on, delighted thus to roam
Through rugged Nature far from friends and home.
She reach'd at last a lone and blest retreat,
A favorite grove, adorn'd by rustic seat,
By blooming rose and vine transplanted there
With taste, that left unchanged its rural air.
The wild clematis twined its whitening globes,
The opening violets spread their purple robes,
The drooping bowers, the vines that clustering hung
Thick wreaths of tendrils pendent boughs among,
And beech and deep-green cedar, wide display'd
In sombre arches, cast around dark shade
So dense the day-beams there but faintly shone,
Mild tinging leaves by other seasons strown.

A foaming stream, that in its blithesome play

Like night-stars dazzled as it dash'd away,

Refresh'd the grove, enlivening every bloom,

And, gently whispering, broke the silent gloom.

IV.

The maiden sat, in loosen'd robes attired,

Her features flush'd, her youthful beauty fired.

The fix'd and changeless gaze her looks disclose,

The pensive mood, the posture of repose,

The eye upraised, the pleasant, cloudless smile,

Almost reveal the thoughts her hours beguile.

What manly form, advancing, cross'd the glade?

His careless steps these quiet scenes invade;

Abruptly turning, with unerring eye

Finds the grove-path and treads it hurriedly.

His handsome features and the noble air,

The mental beauty brightly written there,

His light and lofty bearing, and his dress
In unison with all his looks express,
Proveth, as far as outward sign will show,
That youth hath seldom warr'd with toil or woe.
Kindly received, and standing by her side,
To hurried question he had scarce replied
In guarded language, ere the gentle maid
A strange unquiet in her looks betrayed.
The rose, the color, from her features fled,
A fearful paleness o'er her beauty spread,
Too like the snows a north cold blast will fling
O'er blossoms rich, the first young bloom of Spring.
Impatiently she gazed above, around,
The swaying trees now wake a startling sound.
Another steadfast and more searching glance
Calm and convince her that no foes advance,
And yet how low she breathes each whispered word,
As if perchance her language might be heard!—

"Can it *be* true, and not an idle tale,
   That he hath come—hath even made this vale
   His home and shelter, his secure retreat—
   And blasted honor hath declined to meet?"
"His person, ay, his deeds," replied the youth,
"Are known too well, and that may be the truth.
   Last eve a huntsman, passing noiseless here,
   Beheld a being, and approached so near,
   Though moonlight only, he observed him well,
   And knew the man: he says it was MARCEL!"
"Kind language, CLIFTON, for a friendless ear!
   And uttered, doubtless, to augment my fear;
   And would, indeed, could I those words believe;
   But human vision objects may deceive
   At night, when shadows so perplex the eye
   That one well known might pass, if silent, by.
   Did he accost him?"—"No; he only saw
   The darken'd outlines of the wild outlaw.

Who that hath seen him could forget his form,

His pallid features, and his eye of scorn

That brooks no equal, and reverts its gaze

When friendship, greeting, but a smile displays?

Though true that two long years have pass'd away

Since he left here in foreign lands to stay,

Deem not such love as he reveal'd to thee

Could leave you always undisturb'd near me."

V.

ELFLORA shuddered, and in thoughtful gaze

Recall'd to view the scenes of other days,

Before her beauty far was famed or known,

When none but MARCEL sought her highland home,

And felt the fulness of that quenchless fire

Which lives consuming, though all hopes expire—

When her dear father, with prophetic care,

Had marked the youth and bid his child beware;

With anxious pathos breathed it in his prayer,
Then passed away. His words live keenly there.
In silent anguish mourn the fatal hour
Thy beauty's artless art displayed such power!
Keen be thy sorrow, pure and lonely one!
To parent's will thy mind could not succumb.

VI.

"I blame not others," she at last replied;
"The imprudence mine, whatever ills betide.
So unsuspecting in my younger days,
I deemed him worthy, spoke of him with praise,
Accepted favors, and with him alone
In pathless forests rambled far from home,
And, playful, artless, leaned upon his arm.
That kindness, I presume, did all the harm.
Though social with him, *familiar*? NEVER!
I cherish'd only what my will could sever—

A childish friendship, form'd in foolish pride,
Pleased with devotion, careless if denied,
Coldness toward me, a smile to others near,
Disturb'd me not, nor caused a sigh nor tear.
(O hallowed moments in life's golden spring!
And, now remembered, mournful sadness bring.)
At last faint whispers, startling, met my ear
Of deeds unmanly in MARCEL'S career.
Instant, with firmness, I repulsed him then,
Nor him would see, nor sentence from his pen.
And yet, reluctant to believe him base,
Our worthy pastor tried all means to trace,
At my request, that censure to its source,
But failed, nor found in him the least remorse.
So always cautious, so composed with man,
No looks betray him; and, howe'er you scan
His marble visage, none can truly tell
Its thoughtful import, or be calm so well

When prying converse, artful in its aim,
The dark suspicion mingles with his name."

VII.

"Dwell not," said CLIFTON, "on a theme to thee
So fraught with sadness, and so scorned by me;
The leafless Winter, with its frozen sky
Of lowering clouds, and tempests howling by,
Hath passed away, and blooming Spring once more
With life and verdure decks the landscape o'er.
Like it, ELFLORA, rouse thy joyous mind,
What can be hopeless to a heart resigned
And sternly trustful? Future days shall bring
To thee a calmness—ay, an inward Spring.
The little violet, nestling on the rock,
Will bloom secure when forests feel the shock
Of strong tornadoes, and their giant frames
In shattered fragments strew unshaded plains;

Its form so fragile will outlive the storm,
And for that shelter long those cliffs adorn.
Come, the violet imitate, and bless my home,
My pleasant mansion, now, alas! too lone,
And ever will be till thine eye and smile
Illume its halls and all my hours beguile.
Give me the right to sympathize with thee
In sickness and in sorrow, and to be
Thy solace in all trying scenes of life,
Thy dearest shield in MARCEL's subtle strife;
And thou shalt be, in all our walks of joy,
A happy child, and I thy darling boy."
A moment coloring, she returned his gaze,
That steady look a sincere love betrays;
That glance, peculiar to her own dark eye,
Most eloquent, proclaimed her heart's reply;
Through loosened ringlets, clustering round her face,
A grateful tear the ardent youth could trace,

And, half caressing as he clasped her hand,
He seemed by gratitude almost unmanned.

VIII.

"Ills come with time," said he, "then why delay
Our promised union till some distant day?
The threatened danger that enthralls thee now
Argues this eve should consummate thy vow.
To-night, with HOWARD, I will seek thy home;
Our reverend sire shall witness it alone."
"To-night? to-night?" exclaimed the startled maid.
"An only parent claims my love, my aid;
Our union now she firmly would oppose;
I know her thoughts; thy wish must not disclose.
'Twould pain her feelings, agonize her mind,
Pour fire on wounds my love, my words, should bind.
Yet thou art favored; only wait, I pray,
For her dear sake, a more propitious day.

Peaceful while single—death if e'er I wed,

Was EDWARD's threat: that threat is now her dread."

"Never," said CLIFTON, "will the villain dare

To mar a happiness he cannot share.

This eve our nuptials must take place; and now

I ask but this, thy sanction to that vow;

And trust my wisdom to arrange a plan

That none shall know except the holy man.

Rememberest thou the old majestic ash

Whose boughs were shivered by the lightning's flash,

And standing now in ruins, lonely, high,

With dead-leafed arms outstretched across the sky?

To-night in secret to that shelter come;

A moon will fill the heavens when day is done;

Our pastor's quiet home thou knowest is near,

And he will welcome us with friendly cheer.

His aid and service we can always claim;

The marriage from the world he will retain."

IX.

Pleasing to themselves, all Nature pleasing,
ELFLORA and her friend the shades are leaving,
The sunless grove; no living object near
Save songsters chattering to each loved compeer—
The wren, the jay, the brownly-spotted thrush,
The warblers dyed like evening's parting flush—
Enliven still Seclusion's favorite place,
Some twittering joyous; others gayly chase
Their little comrades on from limb to limb,
Or through the air, vociferous, lightly skim,
Till life and love, re-echoed, fill the wood
Which man miscalls a lonely solitude.
But mark! the birds are dashing far and fleet,
The leaves, the air, with rapid wings they beat,
And skyward soar affrighted wildly, shriek.
What being roused them in that calm retreat?

In creviced rock who dare, leaf-shrouded, lay,

Like warrior armed, perchance prepared to slay?

Can it be true? It is indeed MARCEL

Who rises there, where naught of ill should dwell.

His haughty features, broadening to a sneer,

Disclose a spirit warped by wild career,

Maddened by love, by disappointment scarred,

Hardened by crime, by midnight revel marred.

His kindling wrath repulsive looks enhance

To grimness stern, as scornful scowls his glance,

As muttered words reveal the sullen joy

The purposed vengeance which his thoughts employ:

"I too remember the majestic ash

Shattered and shivered by the lightning's flash,

And by that Power, by Heaven, I swear

To-night, poor orphan, I will meet you there—

By subtle arts, if not by merit, gain

Your heart and hand, perhaps your wealth obtain;

And while your soul, incarnate, deigns to be,

No human might shall wrench your form from me.

But come, come, evening; passion's storm, sleep calm;

Triumphant conquest soon shall bring me balm;

My plan, projected to ensnare the maid,

Though baffled now, the coming night will aid."

He ceased, and like the silent, single cloud

Which flings o'er heaven's blue its blackening shroud,

Passed down the glen, and slowly moves away,

And in that wood there seems a brighter day.

X.

ELFLORA'S mansion stands in quietude;

No foes disturb, no noisy guests intrude;

The windows, raised, admit the shine of day,

And glass and massive carved-work catch the ray,

Reflect it o'er the walls, diffuse, unite,

Till all the room with yellow beams is light.

CANTO THE FIRST.

The opening rosebuds, late arranged with care,
And watered fresh, with fragrance fill the air,
And, scattering sweetness, woo the wandering bees,
Which come, and, pilfering, wave those tiny trees
Like gentle zephyrs; and that motion's all
Which stirs within the maiden's lonely hall.
Once more she comes, and noiseless crossed the room,
A moment viewed the buds' unfolding bloom,
A moment grasped a favorite volume nigh,
Fluttered its leaves, then tossed the treasure by;
Then seized a crayon, plucked a pencilled rose
She partly shaded ere last eve's repose,
Contracts her brow, and bending fixed her look;
But dropped the gem, no quietude could brook.
Her thoughts were music, floating fast and free,
Wild warbling forth in tenderest melody.
But, hark! A voice hath breathed ELFLORA's name.
How sudden still, how marble-like, that frame!

A pure regard, a troubled, anxious air,
And pale solicitude's expression there;
Now, winged by love, her steps to chamber glide,
And kneels her by an aged mother's side.
O youth! how envied in thy heart's career
When words shall greet thee half as fond and dear
As those sweet breathings of that filial maid
Which to her parent warmly proffered aid!
That parent! Oh behold her faded eye
Suffused with tears, affection's mute reply,
And hand, slight trembling, on her daughter's hair
In kind approval of her pious care!
Those features, furrowed by the lapse of time,
Show youthful sweetness now in deep decline;
So light and gently fell the power of years
That only in her form their force appears.
What hallowed fortune blessed her troubled day
To feel not weary as it passed away?

Love, ever constant, from her only child,
Her sorrows soothed and all her cares beguiled,
And made her happy. Though almost alone,
A calm contentment always marks her home,
In close communion with one spotless heart
Has peace no other ties could e'er impart.

XI.

"My child, a pure and deep regard for thee
Solicits care. My mind, from fears not free,
Dwells painful on a troubled slumber's dream.
And oh how awful, how confused, that scene!
Strange sounds, strange voices, broke upon my ear;
Beings unknown, but oh not thou, wert here.
I called thy name, I watched the opening door;
Others then entered, but thou cam'st no more.
How still, how lonely, everything appears!
Now speak, my child; can these be idle fears?

Tell me thou wilt not leave me here alone;
Forego thy visit, and this night at home
Remain with me, my only cherished dear;
But two days since the HOWARDS all were here.
What cause, what motive, can impel thy mind,
When I against it am so much inclined?"
"My dearest mother, grieve not thus; we know
Consent thou gavest scarce one hour ago.
The dreamy slumber of this afternoon
Should not, dear mother, fill thy heart with gloom.
All dreams are shadows, to thee however bad,
And life without them brings enough that's sad.
How short thy slumber! See, the cloudless sun
Is shining here as when thy rest begun;
Be tranquil now, and court more sweet repose,
And I will watch till sleep thine eyelids close.
Rest is essential to thy weary mind;
Composure calm will make thee more resigned."

"My own, my daughter, thou hast ne'er before

Beyond my slightest wish e'er ventured more.

So kind and gentle, every act and prayer

Seemed interested in thy mother's care;

Yet now my vital welfare claims regard;

My counsel canst thou thus this day discard?"

ELFLORA listened with profound regret;

In tones of mild and soothing language met

The prudent doubts maternal fondness gave,

And tried by every means her fears to waive.

And yet she told not of that nuptial deed

The night would witness should her hopes succeed;

But tenderly with love, and doubly kind,

And to her wish appearing all resign'd,

With woman's tact and art she gently drew

A full consent to all she wished to do;

Except the marriage, which *must not* be known

Beyond the study of her pastor's home.

CANTO THE SECOND.

I.

WHILE yonder sun illumes with level ray,
While robes of gold the grand old hills display,
The shadows lengthening on the plains extend,
The cooler airs with day diffusely blend,
The towering clouds in crimson mountains rise,
And, 'sembling Etnas, flame one half the skies.
But now they change, the shifting vapors fade;
O'er all the landscape fast descends their shade;
A moment sparkling gleams the setting sun,
And sinking now, the halcyon day is done.
The stars are brightening in the blue above,
And beautify the walks endeared to love;
The songs are hushed; the vales and hills in view,
By evening shadowed, pale their verdant hue;

The air, reposing, yields its ardent light,
The full-orbed moon in glory forms the night.

II.

Through pathless fields, by faithful woodman led,
By branching thorn, by wild-rose thickly spread,
By cliffs and scattered rocks, o'er whispering rills,
By fragrant groves, that gloom the silent hills,
Now turning, and now bending to evade
The drooping limbs 'neath which her childhood played,
ELFLORA reached the plain, and saw the tree,
Dark, standing lone like sail-reefed ship at sea.
The feeble outlines of a human form
Cast a pale figure far across the lawn,
And moving stealthy, sheltered in the shade
The huge trunk in silver moonlight made.
The bride beheld it, and with waving hand
And gentle whisper, stern though sweetly bland,

Dismissed her guidesman, and pursued her way

Athwart the plain, whose paths seemed lit by day,

So bright the verdure that around her lay,

So cloudless clear the orb's refulgent ray.

Now on that person whom the light revealed,

But whose disguise a cunning foe concealed,

She leans reposing, and her quiet air

Betrays no fear that CLIFTON is not there.

And they are silent, moving toward the wood;

But why—why looks she not where late they stood?

Confiding innocent! mistrustless one!

Believing all her sorrows nearly done,

Lightly and buoyant down the sloping way

She glides like snow that falls on fire to lay.

III.

The forest verge is gained, and hidden now

By shadows flung from many a massive bough,

They pause. MARCEL kneels down with searching eye,
Anxious to scrutinize all objects nigh;
A figure, CLIFTON-like, is dimly seen,
Approaching fast where they had lately been—
Not recognized, but moving in faint view
Between him and the pale horizon's hue.
He fancied that dark form can only be
The one from whom it would be wise to flee,
And caught the maiden's arm to lead her on,
Sought the dense shades, and noiseless passed along,
Till now through tangled wood he wends their course,
Down a steep bluff, through rustling brambles force,
Sliding, sustained by limbs and shrubs around,
Some bending breakless, some uproot the ground,
Endangering by the worthless aid they lend,
Like counsel false obtained from faithless friend.
Stern Nature's gloomy castles now are near,
Cliffs, grandly towering above cliffs, appear,

With laurel interspersed, and pillared firm
By giant trees; the vine and woodland fern
Spring greenly there amidst the grayer moss,
And wreathed in folds the massive crags emboss.
Once more they paused. A cavern's secret door
Is seen, when logs and stone that closed it o'er
MARCEL removed, and part exposed to view,
In rock, the dismal entrance riven through.
ELFLORA, trembling, instant turned aside,
Averse to enter e'en with favorite guide;
And when he urged her on against her will,
She felt through every nerve a shuddering chill,
And backward stept, alone some course to take,
And gazed to see what thence might be her fate.
While undecided still her guide led on;
She sighed, and, yielding, followed, and they're gone
In the dark cave which penetrates the hill.
Now rough, now smooth their walk; they stand; all's still.

Then through the cavern instant flashed a light,

And MARCEL's features burst upon her sight.

Wildly she shrieked, and fell, but, falling, found

An arm of strength entwined her waist around,

And laid her lightly on a pallet near

Of otter furs outspread on skins of deer.

IV.

The youth a moment stood, then turned away,

And left the maiden; motionless she lay,

Not breathing, or so gently that her breath

Scarce told that fearful stillness was not death.

But soon life's fountain redly flushed again;

The lips and features moved; some color came;

And, slowly rising, now a single hand

Supports that frame almost too shocked to stand.

Surprised, bewildered, far around she gazed.

What is that scene that thus perplex'd, amazed?

The torch refulgent sheds a golden light,

On every object gleams intensely bright,

And pictures forth the aisle, the dome, the shrine,

Their cast of grandeur, every carved outline,

And pillared halls whose lofty roof o'ershrouds

With gorgeous arches, 'sembling silver'd clouds

Of massive volume seen in summer sky,

When piled in banks the radiant vapors lie;

The curtains rich, while in their breezy flow

Seem changed to stone and hang like drifts of snow,

Reflecting broad, in star-like lustre there,

The flame, whose brightness glows beside the fair.

The spacious cavern high and far extends,

And where yon hall the light with darkness blends,

Columns on columns, rising, faintly loom,

Dimly discerned amidst the dusky gloom.

But different to ELFLORA's startled view

Those objects seem—of lifelike form and hue.

MARCEL she deemed was there, and when she saw
Him thus disguised, and every wild outlaw
That followed him, and form'd his reckless band,
And ghost-like glimmering, and around him stand,
No CLIFTON there, and no chivalric arm
Near to protect, and proud to shield her form,
Tears came, and language of despair intrude,
Which proved her firmness more than half subdued.

V.

A sounding tread now echo'd in the rear,
And from without, advancing fast and near,
The villain comes; surprised, the maiden sees
No demon frowns, but smiles awaked to please.
"Fear not," said he; "though in this lonely cave,
Thou hast a soul to make MARCEL thy slave.
I brought thee hither, anxious to reveal
One truth my heart, though iron, must ever feel.

Behold my features! Ay, look on me now:

My callous feelings cloud my youthful brow.

Despised and spurn'd by one I deem'd my friend,

Reckless of life and careless of my end,

A slander'd victim, sentenced though unheard,

And thou—how faithless!—ne'er repelled that word.

I am no villain, am not even bad,

And comrades vile—great God!—I never had;

So, pure in thought, I am prepared to die;

Yet this to thee, thy friends will dare deny;

They have o'erreach'd me, have estranged thy heart;

But sleepless cunning now unfolds her art,

For thou art captive, and thy friends shall hear

The shout of conquest and behold my sneer!"

At this ELFLORA, with indignant pride,

And rising full, with frowning glance replied:

"MARCEL, that language, breathed to give me pain,

Forbear, ungrateful! Do I merit blame?

Do not my feelings, all my actions, show

A sister's sorrow? Darest thou answer, 'No'?

If that be fiction once against thee hurled,

With truth, poor coward, undeceive the world;

Retrieve thy honor; prove thy motions pure;

Assume that station talents can secure,

And teach mankind one noble lesson yet—

Thy clouded star shall not in darkness set.

Away with folly! cease of love to dream!

Thy name, dishonor'd, go, once more redeem."

VI.

"ELFLORA, *dearest* (frown not at the word;

Thou shalt be mine, the time not long deferr'd),

However glorious, however grand,

To wake the soul and bid it proudly stand

Among the immortals ever known,

A god in mind that time can ne'er dethrone,

All would be naught, a nothingness, to me,
A lifeless chaos, if unblest by thee.
The dark-eyed daughters of chivalric Spain,
And all of Grecia's honor'd old domain,
The maidens of Italia's rosy land,
The dames who tread Circassia's lofty strand,
The noble Briton, and the Frank less free,
Whose stars of beauty burn beyond the sea,
Do not possess, in form nor earthly shrine,
A soul whose flashes lighten brows like thine."

VII.

"Enough! enough! for all that thou canst say,"
Exclaim'd the maiden as she turned away,
"Shall not avail thee, never change my mind,
Though servile praises be with force combined.
I do disdain thee, and I fear thee not;
I scorn thy homage, I despise this plot.

Stand back! hands off! I am not in thy power!
Alone I am not in this trying hour.
In God I trust: I know that he is *here*,
And will protect me if I have no fear;
For ever paralyzed that hand shall be
If laid with passion's dark intent on me."

VIII.

The villain paled, then slowly paced the room,
Now near the light, and now in distant gloom,
And, frowning sullen, gazed upon the floor,
The haughty maiden, and the cavern-door,
As though his feelings held wild war within,
Now swayed by terror, now beguiled by sin.
While thus his passions jarr'd in deep contrive,
He left the prisoner, buried, yet alive;
But, pausing, closed the cave with artful care,
Then gazed till satisfied no spy was there;

Then, muttering to himself, he forced his way
In that straight course he knew her mansion lay.
Short was the time till on his eager view
Light fill'd the windows, brighten'd, then withdrew,
As though the inmates pass'd from room to room
With all that hurry caused by mental gloom.
Near now, and almost there, he plainly saw
They talk'd perchance of him, the wild outlaw;
He reck'd not what they said, but boldly sprung
Up the old steps and stood that crowd among,
Which widen'd speechless, and with steady eye
Watch'd his, that flash'd a moment to defy,
Then gather'd quickly to a smile of scorn,
Which proved at once with whom the bride had gone.
Soon round the culprit press'd the closing crowd,
And silence broke in voices censuring loud,
And, waking justice roused in every breast,
They soon, with iron hands his form arrest,

With cords immediate tensely bind each limb,

With prudent forethought weapons wrest from him,

And, over-cautious, as they pass'd him by,

Kept a close watch upon his hand and eye,

And barr'd the doors and closed the shutters to,

Lest his associates might the broil renew.

So strange, so startling, all his late career,

So wrapt in mystery, so involved the sphere

In which he triumph'd since exiled from home,

That wild conjecture filled the void unknown,

And whisper'd ominous to condemn MARCEL:

"Behold his forehead, gash'd by sword or shell,

And mark his bearing, watch his kindling eye,

When cannon-thunder booms along the sky."

IX.

The prostrate captive, though disarm'd and bound,

Now raised his eye and quiet gazed around,

Until his searching glance met CLIFTON'S; then
A shout and laugh which shock'd the crowd of men
Broke from his lip, and told at last too well
No threats could conquer and no arm could quell,
Unless they silenced him by sudden death,
Or stifled with strong hand his venom'd breath.
"You have my weapons," then he coolly said;
"Now stab my heart or shoot me through the head;
Commit at once the wished-for homicide.
Why stand reluctant? Can it kill the bride?
Ay, dare to do it! She shall feel the blow,
For where she slumbers you shall never know.
She is imprisoned in a secret cave
Where not a hand but mine her life can save;
No food to nourish, and no water near,
She dies if these mean cords detain me here.
But not unconscious will she yield her breath:
Old Time, remorseless, shall behold stern Death

Deep waste her form and wear her strength away,
Till 'feebled Nature scarce hath power to pray;
Then loathsome vermin will from creviced wall
Steal forth, and coldly o'er her features crawl,
And she shall feel them, try to scare in vain;
They undisturbed will stay, and, gnawing, pain.
Unloose me now, or by that cave I swear
As now she shall remain, shall perish there!"

X.

The pause which followed open'd to the ear
Deep groans of anguish in the chamber near,
So rending to the heart all rush'd within
And saw the ruin wrought by giant Sin.
They loosed MARCEL, and, holding, brought him nigh;
That sight which fill'd with tears each manly eye
He coldly gazed upon, perchance with hate,
Believing she half caused his gloomy fate.

Raised on her couch by those who near her stand,
The feeble being stretched her fleshless hand
And grasped MARCEL'S, and, trembling, mildly said,
" Can this be thee ? is all thy virtue fled ?
Oh no, it is not. Give me back my child!
Oh, EDWARD! is thy *manhood* all exiled ?
Too often thou hast seen her soothing care,
How dear she was, how warmly she would share
The little toils my daily comforts ask
From those who deem'd their watch a weary task ;
Night after night, all-anxious, near me stay
When servants, coldly kind, would steal away,
Regardless of their charge, and only come
When gentler hands their little rites had done.
It seems but yesterday—dear happy hours !—
Her infant form in grove and garden-bowers
Play'd round me, artless, wild with joyous glee,
Careless of all care, from every sorrow free,

Till all at once my little infant grew

The guardian angel hopeful Fancy drew.

And must I lose her now? is such pure worth

Rewarded only by the thorns of earth?

Be generous, please! I ask it with a prayer,

And bless the hand that will my daughter spare."

XI.

In thoughtful silence MARCEL turned aside,

And from the chamber stept with humbled pride,

And by a window sat, and propp'd his hand

Against his brow, whilst others, gazing, stand.

Then HOWARD, keenly anxious to restrain

The storm of passion and the bride regain,

Now spoke so kindly to that shattered heart,

In tones meek goodness only can impart,

That he was summoned to be seated nigh

By meaning glance of EDWARD's shaded eye.

"My friend and pastor," then replied MARCEL,
"Thy words revive me; thou hast counselled well.
Now to ELFLORA come, and feel no fear;
By daylight, I presume, she can be here.
It is her wish, to-night, to have thee there,
And I alone to her must not repair.
Do not, I pray, decline her last appeal
When life demands it; this thou canst but feel.
I must acknowledge that I could not see
ELFLORA'S parent when she gazed on me,
But my poor mother rose before my eye
In that last sleep in which I saw her die.
It hath unmanned me, hath so chilled my frame
A deathlike faintness now unnerves my brain.
I ask one moment of my early years;
My deep-wrought feelings need relief in tears,
But they refuse my heart that calm repose
So pure, so grateful, when the tear-drop flows."

He ceased, and rose to leave and lead the way,
But voice imploring now prolongs his stay,
Called to the chamber; HOWARD stands alone,
With CLIFTON near him, who in whispered tone
Talk of the danger of their absent friend,
And offer plans that might perchance extend
Relief, though transient, till the early dawn
Revealed to them the course the bride had gone.
"What fatal rashness," then the pastor said,
"To be by him to that lone cavern led!
For well I understand his dark design;
But wisdom may suggest to not decline;
There might some unexpected means arise,
Some unforeseen event, to aid surprise.
Perhaps, like huntsmen, I could mark the track;
A limb snapped here and there, while wandering back,
Among the leaves would whiten in the sun,
And you that splintered path might find, and come."

Then CLIFTON answer'd, "Do, my noble friend,
Accompany thus; by every means extend
Relief and comfort to that spotless love,
And soon as morning dawns yon hills above
I'll rush to save you, search and find the cave,
With strong associates, comrades true and brave."
The hall with others EDWARD enter'd now;
Some change seem'd wrought upon his iron brow;
But, whispering sternly in his rival's ear,
He utter'd words that HOWARD must not hear:
"Return my weapons! Dare to watch my path,
The bride shall famish, you shall feel my wrath."

XII.

The purest hopes induced the aged sire
The offer to accept; while they retire
Despair and sorrow paint with deeper gloom
The looks of those who linger in that room,

Inactive, cautious, lest they might defeat
By steps imprudent plans with risks replete.
One spoke regretful that they did not slay,
Or keep the culprit firmly chain'd till day,
And boldly promised to uproot the wilds,
The hills and valleys, and all deep defiles,
But he would find her, and securely save
From demon hands, and bear her from the cave.
Others, excited less, made calm reply:
"While that was being done the bride would die."
"Then why not force the robber to reveal
The place by burning brand or bleeding steel?
His poniard's point I would have gently prest,
Then harder, still more hard, against his breast,
And threaten'd to the hilt to plunge the blade;
We might, all doubtless, thus have gained the maid."
But one acquainted long with stern MARCEL,
Said, "Mildness, and not force, can only quell."

While thus too loudly flow'd each useless word,

Within the chamber dying groans were heard,

And those who coldly watch'd the glazing eye

Whisper, " Ere dawns the day, think not she'll die?"

CANTO THE THIRD.

I.

ONCE more we leave the mansion and its guests,
And in that cave in which ELFLORA rests
Behold her prostrate, slumbering, all alone,
Her hair neglected, furs half round her thrown,
The tears undried, the frowning features pale,
The wreck of dreams—the dreams that still prevail.
The lips, slow moving, mark a troubled sleep,
She sighs, again is wrapt in slumbers deep,
And now more loud, more heavy, breathing makes,
And now, in one convulsive start, awakes,
Looks round her with a wild, distracted air,
A moment silent, and then kneels in prayer,
While o'er her forehead droop dark locks of hair.

Her trustful language, breathed in mellow'd tone,
Falls not unheard; she is not now alone :
Unnoticed by ELFLORA, softly come
Her friends, now resting till the prayer is done.

II.

The maiden rose, and dried the grateful tear
Which fill'd her eye, and gazed, but not with fear,
Upon those forms approaching near her now,
Though one had malice stamp'd upon his brow.
Her lofty spirit had resign'd its fate,
And stood serenely calm'd, resolved to wait
With unchanged mind the dark, uncertain doom,
Which hope, though nightless, scarcely could illume.
One glance of kindness on her friend she threw,
But in that look no smile beam'd forth to view;
The settled stillness of her feeling heart
To features snowy pale no gleams impart

Of that warm welcome which in early days
Oft wing'd her step along the flowery ways,
When in the forest she beheld his form
Coming like sunlight through an evening storm.

III.

With sullen calmness, glancing at the bride,
MARCEL in silence proudly by her side
Now placed himself, and grasp'd her trembling hand,
And cool, though earnest, issued his command
To HOWARD, standing mute in meekness by,
But who the mandate heard without reply.
"My friend," said he, "I brought thee here to-night
With honest intent and a sense of right,
That by thine office thou may'st end the strife
Which threatens now an aged parent's life.
Now, marry us at once, and legally too;
Speak out! do quickly what thou hast to do!

Or *die* in darkness, starved within this cave,
And all alone, where not a hand can save,
Where all thy time untold shall pass away,
Where never more shalt thou behold the day.
The task is easy; what we both require,
This reconciling act, is her desire;
This sacrifice she is prepared to make,
This very hour, for her dear mother's sake.—
Silence, ELFLORA! Wilt thou dare decline?
What must be *must be*, and thou shalt be mine!
When we are married, and from CLIFTON free,
A model husband I intend to be."
The youth, resolving that the nuptial rite
Should be accomplish'd ere the short-lived night
Could lift its mantle and reveal the den
To search untiring of relentless men,
Now aimed his weapon at the pastor's.breast,
And, grimly frowning, urged his dark request

CANTO THE THIRD.

With such wild gestures, such o'erwhelming scorn,
Few could withstand it or live out the storm.
Yet HOWARD was calm; the mildness seen before
The aged pastor's prayerful features wore
As he, unconscious, seemed to pass away
In thought and spirit to those realms of day
Where saints are resting, and where never cease
That life whose pastime mars no holy peace.
The bride, with eye on his, hath inly caught
The inspiring tone his trustful feelings wrought,
And stands unshaken, with an air and mien
Of stern composure equal to the scene,
And loosed at once the hand that held her arm,
And lean'd, though lightly, on her pastor's form.
Once more the ruffian, roused to foaming wrath,
Satanic fury hurls 'round Beauty's path;
Yet, undisturb'd, the maiden dared to smile,
Sustain'd by steadfast faith against his guile.

The villain, trembling, now approach'd more near;
His arms re-priming to excite their fear,
A moment on them drew unerring aim,
Then, sharply wheeling, at the torchlight flame
Discharged the weapon, and the shatter'd fire
To cinders flash'd; and as the sparks expire
A sudden darkness, deeper than the gloom
Of midnight tempests when there shines no moon,
Pervades the cavern, render'd doubly drear
By echoes far roll'd back from echoes near,
Like bolts of thunder shaking Summer's sky,
Or crash of woods when winds are howling by.

IV.

MARCEL advanced and clasp'd around the waist
The startled beauty, and with caution traced
Those gloomy chambers to the distant door,
And moved so noiseless as the bride he bore,

CANTO THE THIRD.

Emerging outward, that the aged sire
All guidance lost as fast their steps retire,
And he remains immured within the cave—
His home in future, soon perchance his grave.
And now the cavern firmly once again
Is closed; unopened it may˙long remain.
And this MARCEL accomplished with a care
That filled ELFLORA's heart with wild despair.
Once more his arm encircled hers with force,
And downward toward the stream he turns their course,
Through woods so rugged that the flying deer
Would pause for shelter there, and feel no fear.
The boughs above them and the cliffs around,
The lonely hour, the silence so profound
(Save the monotonous and constant roar
The river sent in thunder to the shore),
Fell on the maiden's heart, and swept away
A transient moment all the ills of day;

But ever and anon they came in view,
And left upon her brow an icy dew.
Yet hope of rescue half sustains the bride,
And cautious glances cast from side to side
By him who leads her plainly now reveal
He fears those riven crags her friends conceal.
With weapon ready and with noiseless tread,
With hand on hers, he gently leans his head,
And, often pausing, with nice care surveys
Each moving bramble which the light wind sways.
Old trunks of shatter'd trees, that branchless stood
Like spectres peering through the darksome wood,
Proud oaks, prostrate by the power of Time,
High rocks, grotesque and capped with creeping vine,
Seem living forms by troubled Fancy's aid
As scatter'd moonlight fitful on them play'd.
But, ever watchful, with a steady hand
To execute his selfish heart's command,

He moves as fearless to the river-shore
As ever hero Freedom's standard bore,
Though in the forest, in each darken'd glen,
His guilty conscience sees unflinching men,
Whose faithful rifle in the deadly strife
Leaves nothing living when it aims at life.
At last in safety on a giant rock
That turns the stream, nor trembles at the shock,
They stand, and, speechless, far around survey
The rough, wild scene, and wilder waters' play,
For there the Susquehanna rolls along
A boisterous current, clear yet pouring strong.

V.

The youth now shoved in view an oarless boat,
And moor'd it near them, on the surge afloat,
And instant in it placed the captive bride,
And to her questions strangely calm replied:

"Thy fate, ELFLORA, who shall ever mark?
Alone and friendless in this fragile bark,
I soon shall launch thee on the foaming spray,
Where none can rescue to entomb thy clay,
But hungry fish of thee will make their meal
Ere scarce thy drowning form hath ceased to feel.
And hast thou courage in this hopeless hour
Death thus to meet, and yet possess the power
To change my purpose and establish peace?
Now pledge thy honor, and this plot shall cease,
To love me always, or at least to live
With me as mine, if love thou canst not give.
Dare to refuse me, I will seek thy home
When thou art gone, and by false words alone
Entice young CLIFTON to the secret cave
Where HOWARD lingers my imprisoned slave,
And there, redeemless, with thine aged friend
His joys, his sorrows, and his life shall end.

Thy friendless mother (such she then will be)—
Ay, now thou tremblest, and perchance may see
Some sacrifice to her pure love is due
If thou wouldst not thy hand with blood imbrue."

VI.

Her doom the maiden heard with tearless eye,
And mark'd composure showed in her reply:
" Peace, peace I will not purchase, nor my life,
Nor partial freedom, in this shameful strife
By sacrifice of right, in which I trust,
When conscience tells me it would not be just.
This hour have I no friend to rescue me?
Look round you, and behold in every tree,
· In every leaf, in every flower that blooms,
In every atom that the plant consumes,
In every star, in every ponderous world,
In every ray of light by suns unfurl'd,

In air, in lightning, in the restless sea,
GOD was, and is, and hence shall ever be,
Owning all space, pervading every sphere;
And, to my heart, so ever sheltering near.
MARCEL, my soul is safe: I have no fear;
My trust, my faith in GOD, will save me here.
If I must perish, if my hour hath come,
'Tis His high will, not thine, poor wretch! when done."
MARCEL, perceiving by her earnest tone,
Which spoke her feelings more than words had shown,
How vain, how hopeless, every subtle art
To win or conquer that unshaken heart,
Now shoved the vessel, and it dashed away
Athwart the billows, bounding 'midst the spray,
Which rocks opposing burst to sparkling foam
As whirl'd the currents, sidelong swiftly thrown,
Like gliding serpents wreathing round their path,
And backward curling with envenom'd wrath.

The skiff that bore her seem'd a shatter'd speck,
And the wild waters, dark without the wreck,
Appear'd to bear her fast and far away,
As she, apparent, prostrate on them lay.
And as the distance shrouded all from view,
Save mist and vapors form'd for morning's dew,
And waving rapids that roll'd madly on,
The villain, musing, murmur'd, "She is gone!"

VII.

In restless mood he breathed a heavy sigh,
With scanning glance surveyed each object nigh,
Then hastening upward on the hill's incline,
From rock to rock, with hand from limb to vine,
From many a pendent branch accepting aid,
He gained a level crown'd with younger shade,
Some distance from the fatal scene. And now
He leaves the forest, feels upon his brow

The zephyr's balm, which swept the fields and bring
The lavish fragrance shed by rosy Spring.
Through the bright moonlight, on the flower'd lawn,
Near the ash tree, appears a moving form
Engaged intently with a strolling pack
Of faithful dogs that search for some lost track.
With step more cautious EDWARD crossed the plain,
And unperceived upon the huntsman came,
And seized his arm with such an iron grasp
That CLIFTON started with a sudden gasp.

VIII.

"Did I not warn thee, if you sought the cave
Or dared to watch me, there was naught could save
The bride from vengeance, should you chance to find
The place or cavern where she lives enshrined?
Yet mark this effort: here are dogs that know
The tread of maiden from her manly foe,

And now may lead you, like some human guide,
The path, though printless, where I led the bride.
I see the motive, and the project laid
That shout or whistle shall bring to your aid
Intrepid huntsmen, who are even now
Concealed in numbers on yon wooded brow;
I saw them there, and thought that I could trace
The form or features of some well-known face.
They did not see me; nay, and it was well;
I might have sent them howling through the dell,
Or laid them lifeless with my pistol-ball,
Nor left one witness to record their fall.
All this is useless, since I come to crave
That you will join us in the secret cave,
That we together may adjust our cause,
And shield my errors from the outraged laws;
For which I give thee thy unsullied bride,
Still spotless pure, still worthy of thy pride.

But all must pledge me by a solemn oath
A lasting friendship, that I may in both
Thyself and pastor find sufficient aid
The law's stern justice rightly to evade.
The noble HOWARD, ever meaning well,
Hath kindly bound him always to repel
The groundless slanders Envy cast on me,
And asks all friends to sanction his decree."

IX.

"Although imprudent, and perchance unwise,
To credit language which I should despise,
I will to see them venture in the cave,
Yet grant no promise such as that you crave
Before our meeting. And I ought to know
What cause, what circumstance, hath made my foe
So very gentle and his words so bland.
Most certain, something hath thy soul unmanned,

And changed affection for the friendless maid
To selfish interest since she was betrayed."
"There is, and thou shalt hear it," said MARCEL.
"Thou knowest ELFLORA—ay, perchance too well—
The soothing sweetness of her quiet way;
But only rouse her or infringe her sway,
Provoke her feelings, cross her stubborn will,
There's not a panther prowling on yon hill
But I would rather fondle to reclaim
To gentle loving than that fiery dame.
If you such creatures can domesticate,
Come get her, and GOD help thee and thy fate!"

X.

'Twas nearly day as downward through the wood
They bent their course, though every object stood
In all that shadowed loveliness of night
Which rests on earth when planets all are bright—

When clouds are only scatter'd far and few,
Making the clear a purer, deeper blue.
Such was that balmy morn; the setting moon,
Half down the west, was near her mountain-tomb;
The winds were still, the birds not yet in song,
And all was silent, as they moved along
Through forests gilded by the level ray,
Which slept on leaves and cliffs that walled their way.
While slow they walked by crags and giant trees,
There came a sound, so like the whispering breeze
That none but CLIFTON, haply listening, caught
The low-breathed murmur, and its meaning sought
By hurried glance around and through the shade,
Where Night and Silence seemed in slumber laid.
And there a form he saw approaching near
With cautious step, perchance controlled by fear.
The moon a glory round her features threw
As she in breathless quiet nearer drew,

And stood so still, so pale, that Death seemed there;
Nor corse nor statue ever shone more fair.
"Behold that madman!" then exclaimed the bride.
"In oarless vessel, on yon river's tide,
 He dared to launch me from the rugged shore,
 The sport of waters, whose uneven floor
 Too well he knows is formed of countless rock,
 That skiff, unmanned, cannot escape their shock.
 Yet GOD, all potent, turned it in a cove,
 And waved it gently by a pendent grove;
 I grasped a limb, and slowly dragged the boat
 Where still it lies, half grounded, half afloat,
 On a sand-shore, beneath the eagle's tree;
 I left it there, and now once more am free!"

XI.

The villain stood confounded and amazed;
His eye to CLIFTON'S slow and sullen raised,

But sudden turned him from that lightning gaze,
As though he caught the sun's unclouded blaze.
A child, though feeble, now might grasp his hand
And hold him captive, e'en assume command
By cautious language o'er that cowering youth,
So shamed, so withered by the voice of truth,
His brute-like passions were subdued to sleep;
A nerveless tremor o'er him seem'd to creep;
An icy moisture gathered on his brow;
A woman's mildness might control him now
With gentle accents, careful to convey
No word imprudent to prolong the sway.
But CLIFTON carries his protecting arms;
His dirk, his pistol, have peculiar charms,
And why not use them 'gainst his mortal foe?
He wished to conquer; why the chance forego?
He aimed the weapon at his rival's breast,
And shouted, "Stand, sir! I at once arrest!

One single move, one act to rescue, try,

That very moment, dastard wretch! you die!

Kneel at my feet! Obey! I bid you kneel,

Or crashing shot this instant flashing feel!"

XII.

Like rising tempest lit by sheets of fire,

The culprit reddened, and the kindling ire

Curled his proud lip and surged and heaved his frame,

Till from his features flashed a battle-flame.

So startling sudden burst his strength in storm

That CLIFTON scarcely could an act perform

Before the dagger, threatening, o'er him swung

The moment that his weapon aimless rung.

Now breast to breast, and each armed hand to hand,

They strike and swerver, now advance and stand.

Each gleaming poniard drips with smoking blood,

From either bosom spouts the sanguine flood;

That dash a channel carves across the cheek;

The villain bleeds when he essays to speak.

But now a passage of that practised foe

Disarms the other, who but foils each blow

By Nature's weapons and by manly strength;

But, staggering near, his death seems fixed at length.

"This instant, CLIFTON, seals at last thy doom;

To respite thee high Heaven hath no boon,"

Exclaimed the ruffian as he raised his dirk,

With passion trembling, to complete the work.

ELFLORA, where art thou this fearful hour?

For safety dost thou seek some hidden bower?

Ah no! She watches with unfaltering eye

A time to interpose, and, stepping nigh,

Her shawl she opened, every fold outspread,

With dauntless courage wrapped the villain's head.

Down o'er his arms the silken garment droops;

His eyes are dark, his hand makes aimless swoops,

And ere his struggles can remove that veil
A rock's sharp fragment, such as strew the dale,
Young CLIFTON grasped, and then with fatal aim
The hooded culprit stretched upon the plain.
They snatch the cover from his quivering face;
He lies so still he seems in Death's embrace.
Just where the forehead joins his raven hair
The blood is bubbling and the skull is bare;
So wide the gash you may perceive the stone,
So strongly hurled, hath crashed the solid bone.

XIII.

The strife's last echo scarce had lull'd away
Ere three armed huntsmen sprang to join the fray
From out the forest, and behind their train
The peaceful HOWARD, hurrying onward, came,
And hailed them promptly, lest to vengeful deed,
In their excitement, they should dare proceed.

But all was quiet when he reached the scene,
And scarce a movement where so late had been
The wild contention and the strokes for life,
The clash of weapons, and the knife to knife.
The ground was crimsoned, and the bloody grass
Was bruised and matted, crush'd by many a pass
Of feet now fettered by approaching death,
Now strongly struggling in the gasp for breath.

XIV.

Elflora wept, and leaned against the arm
Of him whose counsel and whose manly form
Was now essential to sustain her there,
Her fears to soothe, her painful charge to share.
She told, though briefly, all that late had passed—
The launch, the perils, and her walk when cast
Alone at midnight on the trackless shore—
And all the trials that she nobly bore

CANTO THE THIRD.

To aid them both, and then to foil MARCEL,
Whose purpose, threatening, he had dared to tell;
And now, self-censured, she expressed remorse
That all her efforts seemed of death the source.
"Not thine the fault! Oh no, dear faithful bride!
But love of arms," the sacred sire replied—
"The coward's shield, which virtue casts aside.
Ah me! how often have I warned thy friend
That trust in weapons must in failure end!
What human being could have rescued me
Save EDWARD, whom I could not hope to see?
Who else in time could find that secret cave?
And yet I felt that GOD my life would save.
With strong reliance, with undoubting trust,
That in this life He ne'er forsakes the just,
I sat me down and calm'd to peace my mind,
And to His will with joy my fate resigned;

And then to me was opened exit clear;

It is enough that I am safely here.

And now assistance is by these required,

Once more this subject, if again desired,

I will resume when we have less to do;

Our friend revives; his foe seems better too.—

Kind huntsman, lend thine hand, and with thy aid

I think this youth, with care, may be conveyed

To his own dwelling, while we leave the rest

To wait on MARCEL till his wounds are drest,

And if not able then to walk or stand,

They will of course extend a welcome hand

To bear him home, and then provide with care

The aid and comfort he may need when there."

XV.

Assent to this request was made by all,

And two are left to answer EDWARD's call,

While those with CLIFTON carefully move away,
Nor mark the actions of the men who stay.
One grasped a vine and tensely bound each limb,
The other shook the foe, rough-handling him,
To see if conscience was enough restored
To feel the torture which they held in hoard,
And grimly chuckled to perceive his eye
By glance and motion gave a mute reply;
Then, turning to his comrade, coolly said,
"I fear the prisoner will too soon be dead.
Suppose we lynch him while we know he lives,
And reap the pleasure vengeance always gives
To those whose hatred of such evil ways
Compel the hangman to cut short their days?
Could we be censured should we now embark
His fettered person in the selfsame ark
In which he placed the unoffending bride?
His fate, like hers, *his* god may then decide."

"Ay, that will do, but now 'tis nearly day;
Some one may rescue should he hold his way
Till sunlight; therefore shoulder up that side
The burden; thus between us we'll divide.
Hold fast! grip firmer! There! now close your fist—
The one that's under—and I'll grasp your wrist.
Ay, now we have him; if he will not walk,
His feet may draggle." Thus drawled in their talk
While bearing MARCEL roughly through the wood
To where the vessel, still half grounded, stood.
The captive raised at last his drooping head,
Became collected, and with effort said:
"I know that money hath secured your aid,
You bloody scoundrels of the blackest grade!
Bribed to defame me, you have hourly sought
My life, my ruin, and ye flinched at naught
That might accomplish this, his fixed design.
Ye sons of Judas! born to rot in crime,

Go, get the wages fiends should blush to hold,
The blood-stained guineas of his bribing gold,
And then—then into everlasting fire
Depart, ye cursed! ye can climb no higher!"
The nettled woodsmen answered in reply
That all he said was one ungodly lie,
And they could prove it. But their way was queer:
They cuffed and shook him till the proof was clear;
And then, by this time having reached the boat,
In it they placed him, and when well afloat,
Let go the vessel as it felt the wave,
And as the last long shove to it they gave,
A loud farewell in scorn they sent the foe.
Then died MARCEL? or lived? Who cares to know?

XVI.

The welcome morning breaks broad up the sky;
The birds rejoice; the fluttering leaves reply,

Sprinkling moisture bright as showers of snow,

Which zephyrs scatter as they whirl below;

The orb ascending, warmer beams his light;

The winds are freshening; forests feel their might,

And fling their thousand banners to the breeze;

And clouds, like icebergs of the Arctic seas,

Float the deep blue and proudly bear away,

Till heat congenial fills the home of day.

The cliffs are passed; the openings through the wood

Reveal the mansion, shining where it stood,

With tossing shadows on its eastward side,

And windows sparkling when those shades divide.

The eager huntsmen cross the vale between

With lively voice and step, now tread the green,

And pass the portals of the open door.

The chamber gained, what lay their eyes before?

The parent prostrate! Ay, and cold and dead,

Though hours but few had waned since breathing fled;

And near the couch, with outstretched hand and arm,

With looks of anguish fixed upon that form,

ELFLORA stood, as stands the chiselled rock,

Pallid and tranced by that heartrending shock;

And in her dark and moveless eye appears

A grief too deep for words and too intense for tears.

At last a painful shudder flushed her brain,

And language of remorseful import came:

"No more, no more the object of my care!

Oh this, O Heaven! must I learn to bear?"

XVII.

Day after day the bride, in mournful thought,

Her darkened chamber's deep seclusion sought,

And there in secret oft implored in prayer

That her apparent disregard of care

Might be forgiven, and her watchful grief

Be soothed by hope and find in tears relief.

'Tis done! Her nature ere that season passed
Reaction feels, and peace triumphs at last,
Suffusing calmness o'er her clearing brow,
And where it gloomed a smile is radiant now.
The youth, recovered, meets his blooming bride;
In her old hall they stand, now side by side;
The pastor speaks, proclaims that they are one,
And prays to God to bless the deed now done.
But ere he leaves them, with his calm blue eye
On CLIFTON fixed, to claim for his reply
Attention mute to words to him addressed,
His admonition kind he thus expressed:
"A woman's love! Its home within thy heart,
What angel joys, what peace, it can impart!
In cot or castle it will shine the same,
And yield contentment in its magic reign,
Toil, sickness, sorrow in thine earthly home
Its worth celestial make more purely known.

Then cherish always her whom Nature's laws
Hath bound to thee and made thine own her cause.
'Tis Wisdom's counsel. Addle-brained are they
Whom stranger beauty e'er beguiles astray;
Perfect kindness hath almighty power
On earth, in weary life's eventful hour,
To tranquillize the heart, the human soul,
And calm the passions into firm control.
Oh, then be gentle in thy walk through life;
With kindest accents always greet the wife;
Then she will grow more happy every day,
And smile on trouble till it fades away.
The brightest gem, the grandest gift to man,
Since toil for comfort in this world began,
Is woman—noble, loving, faithful, true,
In all she does, in all she ought to do,
If man by KINDNESS only will impart
That strength which lifts to Heaven her trusting heart."

XVIII.

The Spring departs; the Summer suns are high;
Heat fills the air, and dazzling light the sky;
Seasons on seasons roll their shades away;
Shrubs rise to trees, and trees themselves decay.
The change of Nature and the winds and rains
Relentless come, and scarce one mark remains
Of that old mansion, save a crumbling pile
Of logs, which nourish insects that defile;
But as they moulder down to dust and earth,
Their darkening richness gives luxuriant birth
To briers and brambles and the wreathing vine,
Which, rising fast, around the ruins twine.

www.ingramcontent.com/pod-product-compliance
Lightning Source LLC
Chambersburg PA
CBHW020314090426
42735CB00009B/1336